THIS BOOK BELONGS TO:

CONTACT INFORMATION	
NAME:	
ADDRESS:	
PHONE:	

DEDICATION

This Cattle Record Keeping Book is dedicated to farmers and ranchers who want to monitor and record important information regarding their herd.

You are my inspiration for producing this book and I'm honored to be a part of helping you retain and collect data for your cattle operation.

HOW TO USE THIS BOOK

This Cattle Record Book will help you record, collect, and organize your information in an easy to use format.

Here are examples of information for you to fill in and write the details of your cattle business.

Fill in the following information:

1. Planning Calendar- space to write 12-month planning details

2. Calving Record- details for the calf, cow and sir ID, birthdate, sex, calf ease, calf vigor, birth weight, nursed (yes/no), weaning weight, weaning date, cow BCS, and notes

3. A.I. Bull Breeding Record- Cow ID, date, 1st service, Bull ID, estimate calving date, 2nd service, estimate calving date, and AI tech

4. Cow Production History Card- cow registration #, description (color/breed), cow's sire, dam, birth date, purchase date, sir and dam breed, weaning weight, the purchase price

5. Breeding Record- weight at first service (KGS), age at first service (days), heat dates, date of A.I/Bull service, preg. diagnosis date, date to dry, date due to calf

6. Cow ID#- date of sale or removal, reason, sale weight, the sale price/lb, total value, age at first calving (days), age at conception (days), date calved, calf sex, calving internal (days), notes

7. Medical Log- date, cow ID, medication, dosage, diagnosis, and notes

8. Deworming and Immunizations- date, cow ID, dosage, next due, and notes

9. Cattle Sales Record- date, class of cattle, ID number, weight, the price per pound, the total price

10. Expense Record- Expense Item: date, description, feed, supplies, medical, other, total cost, total expenses, net profit or loss

11. Notes- record information regarding farming supplies, current and future market trends, climate and weather patterns, soil quality, feed management, etc.

PLANNING CALENDAR

JANUARY

FEBRUARY

MARCH

APRIL

MAY

JUNE

PLANNING CALENDAR

JULY

AUGUST

SEPTEMBER

OCTOBER

NOVEMBER

DECEMBER

CALVING RECORD

CALF ID	COW ID	SIRE ID	BIRTH DATE	SEX	CALVING EASE	CALF VIGOR

BIRTH WEIGHT	CALF NURSED (YES /NO)	WEANING WEIGHT	WEANING DATE	COW BCS	NOTES

CALVING RECORD

CALF ID	COW ID	SIRE ID	BIRTH DATE	SEX	CALVING EASE	CALF VIGOR

BIRTH WEIGHT	CALF NURSED (YES /NO)	WEANING WEIGHT	WEANING DATE	COW BCS	NOTES

CALVING RECORD

CALF ID	COW ID	SIRE ID	BIRTH DATE	SEX	CALVING EASE	CALF VIGOR

BIRTH WEIGHT	CALF NURSED (YES /NO)	WEANING WEIGHT	WEANING DATE	COW BCS	NOTES

CALVING RECORD

CALF ID	COW ID	SIRE ID	BIRTH DATE	SEX	CALVING EASE	CALF VIGOR

BIRTH WEIGHT	CALF NURSED (YES /NO)	WEANING WEIGHT	WEANING DATE	COW BCS	NOTES

A.I./BULL BREEDING RECORD

COW ID	DATE	1ST SERVICE	BULL ID	ESTIMATED CALVING DATE

DATE	2ND SERVICE	BULL ID	ESTIMATED CALVING DATE	A.I TECH

A.I./BULL BREEDING RECORD

COW ID	DATE	1ST SERVICE	BULL ID	ESTIMATED CALVING DATE

DATE	2ND SERVICE	BULL ID	ESTIMATED CALVING DATE	A.I TECH

A.I./BULL BREEDING RECORD

COW ID	DATE	1ST SERVICE	BULL ID	ESTIMATED CALVING DATE

DATE	2ND SERVICE	BULL ID	ESTIMATED CALVING DATE	A.I TECH

A.I./BULL BREEDING RECORD

COW ID	DATE	1ST SERVICE	BULL ID	ESTIMATED CALVING DATE

DATE	2ND SERVICE	BULL ID	ESTIMATED CALVING DATE	A.I TECH

COW PRODUCTION HISTORY CARD

COW REGISTRATION #:	
DESCRIPTION (BREED/COLOR)	

COW'S SIRE		SIRE BREED:	
COW'S DAM		DAM BREED	
COW'S BIRTH DATE		WEANING WEIGHT	
PURCHASE DATE		PURCHASE PRICE	

BREEDING RECORD

WEIGHT AT FIRST SERVICE (KGS)			
AGE AT FIRST SERVICE (DAYS)			

HEAT DATES	DATE OF A.I/BULL SERVICE	PREG. DIAGNOSIS DATE	DATE TO DRY	DATE DUE TO CALF

COW ID

DATE OF SALE OR REMOVAL:	
REASON	
SALE WEIGHT	
SALE PRICE/LB.	
TOTAL VALUE	

AGE AT FIRST CALVING (DAYS)				
AGE AT CONCEPTION (DAYS)				
DATE CALVED	CALF SEX	CALF NO.	CALVING INTERNAL (DAYS)	NOTES

COW PRODUCTION HISTORY CARD

COW REGISTRATION #:			
DESCRIPTION (BREED/COLOR)			
COW'S SIRE		SIRE BREED:	
COW'S DAM		DAM BREED	
COW'S BIRTH DATE		WEANING WEIGHT	
PURCHASE DATE		PURCHASE PRICE	

BREEDING RECORD

WEIGHT AT FIRST SERVICE (KGS)				
AGE AT FIRST SERVICE (DAYS)				
HEAT DATES	DATE OF A.I/BULL SERVICE	PREG. DIAGNOSIS DATE	DATE TO DRY	DATE DUE TO CALF

COW ID #	

DATE OF SALE OR REMOVAL:	
REASON	
SALE WEIGHT	
SALE PRICE/LB.	
TOTAL VALUE	

AGE AT FIRST CALVING (DAYS)				
AGE AT CONCEPTION (DAYS)				
DATE CALVED	CALF SEX	CALF NO.	CALVING INTERNAL (DAYS)	NOTES

COW PRODUCTION HISTORY CARD

COW REGISTRATION #:			
DESCRIPTION (BREED/COLOR)			
COW'S SIRE		SIRE BREED:	
COW'S DAM		DAM BREED	
COW'S BIRTH DATE		WEANING WEIGHT	
PURCHASE DATE		PURCHASE PRICE	

BREEDING RECORD

WEIGHT AT FIRST SERVICE (KGS)				
AGE AT FIRST SERVICE (DAYS)				
HEAT DATES	DATE OF A.I/BULL SERVICE	PREG. DIAGNOSIS DATE	DATE TO DRY	DATE DUE TO CALF

COW ID

DATE OF SALE OR REMOVAL:	
REASON	
SALE WEIGHT	
SALE PRICE/LB.	
TOTAL VALUE	

AGE AT FIRST CALVING (DAYS)				
AGE AT CONCEPTION (DAYS)				
DATE CALVED	CALF SEX	CALF NO.	CALVING INTERNAL (DAYS)	NOTES

COW PRODUCTION HISTORY CARD

COW REGISTRATION #:			
DESCRIPTION (BREED/COLOR)			
COW'S SIRE		SIRE BREED:	
COW'S DAM		DAM BREED	
COW'S BIRTH DATE		WEANING WEIGHT	
PURCHASE DATE		PURCHASE PRICE	

BREEDING RECORD

WEIGHT AT FIRST SERVICE (KGS)				
AGE AT FIRST SERVICE (DAYS)				
HEAT DATES	DATE OF A.I/BULL SERVICE	PREG. DIAGNOSIS DATE	DATE TO DRY	DATE DUE TO CALF

COW ID #	

DATE OF SALE OR REMOVAL:	
REASON	
SALE WEIGHT	
SALE PRICE/LB.	
TOTAL VALUE	

AGE AT FIRST CALVING (DAYS)				
AGE AT CONCEPTION (DAYS)				
DATE CALVED	CALF SEX	CALF NO.	CALVING INTERNAL (DAYS)	NOTES

COW PRODUCTION HISTORY CARD

COW REGISTRATION #:			
DESCRIPTION (BREED/COLOR)			
COW'S SIRE		SIRE BREED:	
COW'S DAM		DAM BREED	
COW'S BIRTH DATE		WEANING WEIGHT	
PURCHASE DATE		PURCHASE PRICE	

BREEDING RECORD

WEIGHT AT FIRST SERVICE (KGS)				
AGE AT FIRST SERVICE (DAYS)				
HEAT DATES	DATE OF A.I/BULL SERVICE	PREG. DIAGNOSIS DATE	DATE TO DRY	DATE DUE TO CALF

COW ID

DATE OF SALE OR REMOVAL:	
REASON	
SALE WEIGHT	
SALE PRICE/LB.	
TOTAL VALUE	

AGE AT FIRST CALVING (DAYS)				
AGE AT CONCEPTION (DAYS)				
DATE CALVED	CALF SEX	CALF NO.	CALVING INTERNAL (DAYS)	NOTES

COW PRODUCTION HISTORY CARD

COW REGISTRATION #:	
DESCRIPTION (BREED/COLOR)	

COW'S SIRE		SIRE BREED:	
COW'S DAM		DAM BREED	
COW'S BIRTH DATE		WEANING WEIGHT	
PURCHASE DATE		PURCHASE PRICE	

BREEDING RECORD

WEIGHT AT FIRST SERVICE (KGS)				
AGE AT FIRST SERVICE (DAYS)				
HEAT DATES	DATE OF A.I/BULL SERVICE	PREG. DIAGNOSIS DATE	DATE TO DRY	DATE DUE TO CALF

COW ID

DATE OF SALE OR REMOVAL:	
REASON	
SALE WEIGHT	
SALE PRICE/LB.	
TOTAL VALUE	

AGE AT FIRST CALVING (DAYS)				
AGE AT CONCEPTION (DAYS)				
DATE CALVED	CALF SEX	CALF NO.	CALVING INTERNAL (DAYS)	NOTES

COW PRODUCTION HISTORY CARD

COW REGISTRATION #:			
DESCRIPTION (BREED/COLOR)			
COW'S SIRE		SIRE BREED:	
COW'S DAM		DAM BREED	
COW'S BIRTH DATE		WEANING WEIGHT	
PURCHASE DATE		PURCHASE PRICE	

BREEDING RECORD

WEIGHT AT FIRST SERVICE (KGS)				
AGE AT FIRST SERVICE (DAYS)				
HEAT DATES	DATE OF A.I/BULL SERVICE	PREG. DIAGNOSIS DATE	DATE TO DRY	DATE DUE TO CALF

COW ID #	

DATE OF SALE OR REMOVAL:	
REASON	
SALE WEIGHT	
SALE PRICE/LB.	
TOTAL VALUE	

AGE AT FIRST CALVING (DAYS)				
AGE AT CONCEPTION (DAYS)				
DATE CALVED	CALF SEX	CALF NO.	CALVING INTERNAL (DAYS)	NOTES

COW PRODUCTION HISTORY CARD

COW REGISTRATION #:			
DESCRIPTION (BREED/COLOR)			
COW'S SIRE		SIRE BREED:	
COW'S DAM		DAM BREED	
COW'S BIRTH DATE		WEANING WEIGHT	
PURCHASE DATE		PURCHASE PRICE	

BREEDING RECORD

WEIGHT AT FIRST SERVICE (KGS)				
AGE AT FIRST SERVICE (DAYS)				
HEAT DATES	DATE OF A.I/BULL SERVICE	PREG. DIAGNOSIS DATE	DATE TO DRY	DATE DUE TO CALF

COW ID

DATE OF SALE OR REMOVAL:	
REASON	
SALE WEIGHT	
SALE PRICE/LB.	
TOTAL VALUE	

AGE AT FIRST CALVING (DAYS)				
AGE AT CONCEPTION (DAYS)				
DATE CALVED	CALF SEX	CALF NO.	CALVING INTERNAL (DAYS)	NOTES

COW PRODUCTION HISTORY CARD

COW REGISTRATION #:				
DESCRIPTION (BREED/COLOR)				
COW'S SIRE		SIRE BREED:		
COW'S DAM		DAM BREED		
COW'S BIRTH DATE		WEANING WEIGHT		
PURCHASE DATE		PURCHASE PRICE		

BREEDING RECORD

WEIGHT AT FIRST SERVICE (KGS)			
AGE AT FIRST SERVICE (DAYS)			

HEAT DATES	DATE OF A.I/BULL SERVICE	PREG. DIAGNOSIS DATE	DATE TO DRY	DATE DUE TO CALF

COW ID

DATE OF SALE OR REMOVAL:	
REASON	
SALE WEIGHT	
SALE PRICE/LB.	
TOTAL VALUE	

AGE AT FIRST CALVING (DAYS)				
AGE AT CONCEPTION (DAYS)				
DATE CALVED	CALF SEX	CALF NO.	CALVING INTERNAL (DAYS)	NOTES

COW PRODUCTION HISTORY CARD

COW REGISTRATION #:			
DESCRIPTION (BREED/COLOR)			
COW'S SIRE		SIRE BREED:	
COW'S DAM		DAM BREED	
COW'S BIRTH DATE		WEANING WEIGHT	
PURCHASE DATE		PURCHASE PRICE	

BREEDING RECORD

WEIGHT AT FIRST SERVICE (KGS)				
AGE AT FIRST SERVICE (DAYS)				
HEAT DATES	DATE OF A.I/BULL SERVICE	PREG. DIAGNOSIS DATE	DATE TO DRY	DATE DUE TO CALF

COW ID

DATE OF SALE OR REMOVAL:	
REASON	
SALE WEIGHT	
SALE PRICE/LB.	
TOTAL VALUE	

AGE AT FIRST CALVING (DAYS)				
AGE AT CONCEPTION (DAYS)				
DATE CALVED	CALF SEX	CALF NO.	CALVING INTERNAL (DAYS)	NOTES

COW PRODUCTION HISTORY CARD

COW REGISTRATION #:	
DESCRIPTION (BREED/COLOR)	

COW'S SIRE		SIRE BREED:	
COW'S DAM		DAM BREED	
COW'S BIRTH DATE		WEANING WEIGHT	
PURCHASE DATE		PURCHASE PRICE	

BREEDING RECORD

WEIGHT AT FIRST SERVICE (KGS)			
AGE AT FIRST SERVICE (DAYS)			

HEAT DATES	DATE OF A.I/BULL SERVICE	PREG. DIAGNOSIS DATE	DATE TO DRY	DATE DUE TO CALF

COW ID

DATE OF SALE OR REMOVAL:	
REASON	
SALE WEIGHT	
SALE PRICE/LB.	
TOTAL VALUE	

AGE AT FIRST CALVING (DAYS)				
AGE AT CONCEPTION (DAYS)				
DATE CALVED	CALF SEX	CALF NO.	CALVING INTERNAL (DAYS)	NOTES

COW PRODUCTION HISTORY CARD

COW REGISTRATION #:			
DESCRIPTION (BREED/COLOR)			
COW'S SIRE		SIRE BREED:	
COW'S DAM		DAM BREED	
COW'S BIRTH DATE		WEANING WEIGHT	
PURCHASE DATE		PURCHASE PRICE	

BREEDING RECORD

WEIGHT AT FIRST SERVICE (KGS)			
AGE AT FIRST SERVICE (DAYS)			

HEAT DATES	DATE OF A.I/BULL SERVICE	PREG. DIAGNOSIS DATE	DATE TO DRY	DATE DUE TO CALF

COW ID #	

DATE OF SALE OR REMOVAL:	
REASON	
SALE WEIGHT	
SALE PRICE/LB.	
TOTAL VALUE	

AGE AT FIRST CALVING (DAYS)				
AGE AT CONCEPTION (DAYS)				
DATE CALVED	CALF SEX	CALF NO.	CALVING INTERNAL (DAYS)	NOTES

COW PRODUCTION HISTORY CARD

COW REGISTRATION #:			
DESCRIPTION (BREED/COLOR)			
COW'S SIRE		SIRE BREED:	
COW'S DAM		DAM BREED	
COW'S BIRTH DATE		WEANING WEIGHT	
PURCHASE DATE		PURCHASE PRICE	

BREEDING RECORD

WEIGHT AT FIRST SERVICE (KGS)			
AGE AT FIRST SERVICE (DAYS)			

HEAT DATES	DATE OF A.I/BULL SERVICE	PREG. DIAGNOSIS DATE	DATE TO DRY	DATE DUE TO CALF

COW ID #	

DATE OF SALE OR REMOVAL:	
REASON	
SALE WEIGHT	
SALE PRICE/LB.	
TOTAL VALUE	

AGE AT FIRST CALVING (DAYS)				
AGE AT CONCEPTION (DAYS)				
DATE CALVED	CALF SEX	CALF NO.	CALVING INTERNAL (DAYS)	NOTES

COW PRODUCTION HISTORY CARD

COW REGISTRATION #:			
DESCRIPTION (BREED/COLOR)			
COW'S SIRE		SIRE BREED:	
COW'S DAM		DAM BREED	
COW'S BIRTH DATE		WEANING WEIGHT	
PURCHASE DATE		PURCHASE PRICE	

BREEDING RECORD

WEIGHT AT FIRST SERVICE (KGS)				
AGE AT FIRST SERVICE (DAYS)				
HEAT DATES	DATE OF A.I/BULL SERVICE	PREG. DIAGNOSIS DATE	DATE TO DRY	DATE DUE TO CALF

COW ID

DATE OF SALE OR REMOVAL:	
REASON	
SALE WEIGHT	
SALE PRICE/LB.	
TOTAL VALUE	

AGE AT FIRST CALVING (DAYS)				
AGE AT CONCEPTION (DAYS)				
DATE CALVED	CALF SEX	CALF NO.	CALVING INTERNAL (DAYS)	NOTES

COW PRODUCTION HISTORY CARD

COW REGISTRATION #:			
DESCRIPTION (BREED/COLOR)			
COW'S SIRE		SIRE BREED:	
COW'S DAM		DAM BREED	
COW'S BIRTH DATE		WEANING WEIGHT	
PURCHASE DATE		PURCHASE PRICE	

BREEDING RECORD

WEIGHT AT FIRST SERVICE (KGS)				
AGE AT FIRST SERVICE (DAYS)				
HEAT DATES	DATE OF A.I/BULL SERVICE	PREG. DIAGNOSIS DATE	DATE TO DRY	DATE DUE TO CALF

COW ID #	

DATE OF SALE OR REMOVAL:	
REASON	
SALE WEIGHT	
SALE PRICE/LB.	
TOTAL VALUE	

AGE AT FIRST CALVING (DAYS)				
AGE AT CONCEPTION (DAYS)				
DATE CALVED	CALF SEX	CALF NO.	CALVING INTERNAL (DAYS)	NOTES

COW PRODUCTION HISTORY CARD

COW REGISTRATION #:	
DESCRIPTION (BREED/COLOR)	

COW'S SIRE		SIRE BREED:	
COW'S DAM		DAM BREED	
COW'S BIRTH DATE		WEANING WEIGHT	
PURCHASE DATE		PURCHASE PRICE	

BREEDING RECORD

WEIGHT AT FIRST SERVICE (KGS)			
AGE AT FIRST SERVICE (DAYS)			

HEAT DATES	DATE OF A.I/BULL SERVICE	PREG. DIAGNOSIS DATE	DATE TO DRY	DATE DUE TO CALF

COW ID

DATE OF SALE OR REMOVAL:	
REASON	
SALE WEIGHT	
SALE PRICE/LB.	
TOTAL VALUE	

AGE AT FIRST CALVING (DAYS)				
AGE AT CONCEPTION (DAYS)				
DATE CALVED	CALF SEX	CALF NO.	CALVING INTERNAL (DAYS)	NOTES

COW PRODUCTION HISTORY CARD

COW REGISTRATION #:			
DESCRIPTION (BREED/COLOR)			
COW'S SIRE		SIRE BREED:	
COW'S DAM		DAM BREED	
COW'S BIRTH DATE		WEANING WEIGHT	
PURCHASE DATE		PURCHASE PRICE	

BREEDING RECORD

WEIGHT AT FIRST SERVICE (KGS)				
AGE AT FIRST SERVICE (DAYS)				
HEAT DATES	DATE OF A.I/BULL SERVICE	PREG. DIAGNOSIS DATE	DATE TO DRY	DATE DUE TO CALF

COW ID #	

DATE OF SALE OR REMOVAL:	
REASON	
SALE WEIGHT	
SALE PRICE/LB.	
TOTAL VALUE	

AGE AT FIRST CALVING (DAYS)				
AGE AT CONCEPTION (DAYS)				
DATE CALVED	CALF SEX	CALF NO.	CALVING INTERNAL (DAYS)	NOTES

COW PRODUCTION HISTORY CARD

COW REGISTRATION #:			
DESCRIPTION (BREED/COLOR)			
COW'S SIRE		SIRE BREED:	
COW'S DAM		DAM BREED	
COW'S BIRTH DATE		WEANING WEIGHT	
PURCHASE DATE		PURCHASE PRICE	

BREEDING RECORD

WEIGHT AT FIRST SERVICE (KGS)				
AGE AT FIRST SERVICE (DAYS)				
HEAT DATES	DATE OF A.I/BULL SERVICE	PREG. DIAGNOSIS DATE	DATE TO DRY	DATE DUE TO CALF

COW ID

DATE OF SALE OR REMOVAL:	
REASON	
SALE WEIGHT	
SALE PRICE/LB.	
TOTAL VALUE	

AGE AT FIRST CALVING (DAYS)				
AGE AT CONCEPTION (DAYS)				
DATE CALVED	CALF SEX	CALF NO.	CALVING INTERNAL (DAYS)	NOTES

COW PRODUCTION HISTORY CARD

COW REGISTRATION #:			
DESCRIPTION (BREED/COLOR)			
COW'S SIRE		SIRE BREED:	
COW'S DAM		DAM BREED	
COW'S BIRTH DATE		WEANING WEIGHT	
PURCHASE DATE		PURCHASE PRICE	

BREEDING RECORD

WEIGHT AT FIRST SERVICE (KGS)				
AGE AT FIRST SERVICE (DAYS)				
HEAT DATES	DATE OF A.I/BULL SERVICE	PREG. DIAGNOSIS DATE	DATE TO DRY	DATE DUE TO CALF

COW ID #	

DATE OF SALE OR REMOVAL:	
REASON	
SALE WEIGHT	
SALE PRICE/LB.	
TOTAL VALUE	

AGE AT FIRST CALVING (DAYS)				
AGE AT CONCEPTION (DAYS)				
DATE CALVED	CALF SEX	CALF NO.	CALVING INTERNAL (DAYS)	NOTES

COW PRODUCTION HISTORY CARD

COW REGISTRATION #:			
DESCRIPTION (BREED/COLOR)			
COW'S SIRE		SIRE BREED:	
COW'S DAM		DAM BREED	
COW'S BIRTH DATE		WEANING WEIGHT	
PURCHASE DATE		PURCHASE PRICE	

BREEDING RECORD

WEIGHT AT FIRST SERVICE (KGS)			
AGE AT FIRST SERVICE (DAYS)			

HEAT DATES	DATE OF A.I/BULL SERVICE	PREG. DIAGNOSIS DATE	DATE TO DRY	DATE DUE TO CALF

COW ID

DATE OF SALE OR REMOVAL:	
REASON	
SALE WEIGHT	
SALE PRICE/LB.	
TOTAL VALUE	

AGE AT FIRST CALVING (DAYS)				
AGE AT CONCEPTION (DAYS)				
DATE CALVED	CALF SEX	CALF NO.	CALVING INTERNAL (DAYS)	NOTES

COW PRODUCTION HISTORY CARD

COW REGISTRATION #:			
DESCRIPTION (BREED/COLOR)			
COW'S SIRE		SIRE BREED:	
COW'S DAM		DAM BREED	
COW'S BIRTH DATE		WEANING WEIGHT	
PURCHASE DATE		PURCHASE PRICE	

BREEDING RECORD

WEIGHT AT FIRST SERVICE (KGS)				
AGE AT FIRST SERVICE (DAYS)				
HEAT DATES	DATE OF A.I/BULL SERVICE	PREG. DIAGNOSIS DATE	DATE TO DRY	DATE DUE TO CALF

COW ID #	

DATE OF SALE OR REMOVAL:	
REASON	
SALE WEIGHT	
SALE PRICE/LB.	
TOTAL VALUE	

AGE AT FIRST CALVING (DAYS)				
AGE AT CONCEPTION (DAYS)				
DATE CALVED	CALF SEX	CALF NO.	CALVING INTERNAL (DAYS)	NOTES

COW PRODUCTION HISTORY CARD

COW REGISTRATION #:			
DESCRIPTION (BREED/COLOR)			
COW'S SIRE		SIRE BREED:	
COW'S DAM		DAM BREED	
COW'S BIRTH DATE		WEANING WEIGHT	
PURCHASE DATE		PURCHASE PRICE	

BREEDING RECORD

WEIGHT AT FIRST SERVICE (KGS)				
AGE AT FIRST SERVICE (DAYS)				
HEAT DATES	DATE OF A.I/BULL SERVICE	PREG. DIAGNOSIS DATE	DATE TO DRY	DATE DUE TO CALF

COW ID #	

DATE OF SALE OR REMOVAL:	
REASON	
SALE WEIGHT	
SALE PRICE/LB.	
TOTAL VALUE	

AGE AT FIRST CALVING (DAYS)				
AGE AT CONCEPTION (DAYS)				
DATE CALVED	CALF SEX	CALF NO.	CALVING INTERNAL (DAYS)	NOTES

COW PRODUCTION HISTORY CARD

COW REGISTRATION #:			
DESCRIPTION (BREED/COLOR)			
COW'S SIRE		SIRE BREED:	
COW'S DAM		DAM BREED	
COW'S BIRTH DATE		WEANING WEIGHT	
PURCHASE DATE		PURCHASE PRICE	

BREEDING RECORD

WEIGHT AT FIRST SERVICE (KGS)				
AGE AT FIRST SERVICE (DAYS)				
HEAT DATES	DATE OF A.I/BULL SERVICE	PREG. DIAGNOSIS DATE	DATE TO DRY	DATE DUE TO CALF

COW ID #	

DATE OF SALE OR REMOVAL:	
REASON	
SALE WEIGHT	
SALE PRICE/LB.	
TOTAL VALUE	

AGE AT FIRST CALVING (DAYS)				
AGE AT CONCEPTION (DAYS)				
DATE CALVED	CALF SEX	CALF NO.	CALVING INTERNAL (DAYS)	NOTES

COW PRODUCTION HISTORY CARD

COW REGISTRATION #:			
DESCRIPTION (BREED/COLOR)			
COW'S SIRE		SIRE BREED:	
COW'S DAM		DAM BREED	
COW'S BIRTH DATE		WEANING WEIGHT	
PURCHASE DATE		PURCHASE PRICE	

BREEDING RECORD

WEIGHT AT FIRST SERVICE (KGS)				
AGE AT FIRST SERVICE (DAYS)				
HEAT DATES	DATE OF A.I/BULL SERVICE	PREG. DIAGNOSIS DATE	DATE TO DRY	DATE DUE TO CALF

COW ID #	

DATE OF SALE OR REMOVAL:	
REASON	
SALE WEIGHT	
SALE PRICE/LB.	
TOTAL VALUE	

AGE AT FIRST CALVING (DAYS)				
AGE AT CONCEPTION (DAYS)				
DATE CALVED	CALF SEX	CALF NO.	CALVING INTERNAL (DAYS)	NOTES

COW PRODUCTION HISTORY CARD

COW REGISTRATION #:			
DESCRIPTION (BREED/COLOR)			
COW'S SIRE		SIRE BREED:	
COW'S DAM		DAM BREED	
COW'S BIRTH DATE		WEANING WEIGHT	
PURCHASE DATE		PURCHASE PRICE	

BREEDING RECORD

WEIGHT AT FIRST SERVICE (KGS)				
AGE AT FIRST SERVICE (DAYS)				
HEAT DATES	DATE OF A.I/BULL SERVICE	PREG. DIAGNOSIS DATE	DATE TO DRY	DATE DUE TO CALF

COW ID #	

DATE OF SALE OR REMOVAL:	
REASON	
SALE WEIGHT	
SALE PRICE/LB.	
TOTAL VALUE	

AGE AT FIRST CALVING (DAYS)				
AGE AT CONCEPTION (DAYS)				
DATE CALVED	CALF SEX	CALF NO.	CALVING INTERNAL (DAYS)	NOTES

COW PRODUCTION HISTORY CARD

COW REGISTRATION #:			
DESCRIPTION (BREED/COLOR)			
COW'S SIRE		SIRE BREED:	
COW'S DAM		DAM BREED	
COW'S BIRTH DATE		WEANING WEIGHT	
PURCHASE DATE		PURCHASE PRICE	

BREEDING RECORD

WEIGHT AT FIRST SERVICE (KGS)				
AGE AT FIRST SERVICE (DAYS)				
HEAT DATES	DATE OF A.I/BULL SERVICE	PREG. DIAGNOSIS DATE	DATE TO DRY	DATE DUE TO CALF

COW ID #	

DATE OF SALE OR REMOVAL:	
REASON	
SALE WEIGHT	
SALE PRICE/LB.	
TOTAL VALUE	

AGE AT FIRST CALVING (DAYS)				
AGE AT CONCEPTION (DAYS)				
DATE CALVED	CALF SEX	CALF NO.	CALVING INTERNAL (DAYS)	NOTES

COW PRODUCTION HISTORY CARD

COW REGISTRATION #:			
DESCRIPTION (BREED/COLOR)			
COW'S SIRE		SIRE BREED:	
COW'S DAM		DAM BREED	
COW'S BIRTH DATE		WEANING WEIGHT	
PURCHASE DATE		PURCHASE PRICE	

BREEDING RECORD

WEIGHT AT FIRST SERVICE (KGS)			
AGE AT FIRST SERVICE (DAYS)			

HEAT DATES	DATE OF A.I/BULL SERVICE	PREG. DIAGNOSIS DATE	DATE TO DRY	DATE DUE TO CALF

COW ID

DATE OF SALE OR REMOVAL:	
REASON	
SALE WEIGHT	
SALE PRICE/LB.	
TOTAL VALUE	

AGE AT FIRST CALVING (DAYS)				
AGE AT CONCEPTION (DAYS)				
DATE CALVED	CALF SEX	CALF NO.	CALVING INTERNAL (DAYS)	NOTES

COW PRODUCTION HISTORY CARD

COW REGISTRATION #:			
DESCRIPTION (BREED/COLOR)			
COW'S SIRE		SIRE BREED:	
COW'S DAM		DAM BREED	
COW'S BIRTH DATE		WEANING WEIGHT	
PURCHASE DATE		PURCHASE PRICE	

BREEDING RECORD

WEIGHT AT FIRST SERVICE (KGS)				
AGE AT FIRST SERVICE (DAYS)				
HEAT DATES	DATE OF A.I/BULL SERVICE	PREG. DIAGNOSIS DATE	DATE TO DRY	DATE DUE TO CALF

COW ID

DATE OF SALE OR REMOVAL:	
REASON	
SALE WEIGHT	
SALE PRICE/LB.	
TOTAL VALUE	

AGE AT FIRST CALVING (DAYS)				
AGE AT CONCEPTION (DAYS)				
DATE CALVED	CALF SEX	CALF NO.	CALVING INTERNAL (DAYS)	NOTES

COW PRODUCTION HISTORY CARD

COW REGISTRATION #:			
DESCRIPTION (BREED/COLOR)			
COW'S SIRE		SIRE BREED:	
COW'S DAM		DAM BREED	
COW'S BIRTH DATE		WEANING WEIGHT	
PURCHASE DATE		PURCHASE PRICE	

BREEDING RECORD

WEIGHT AT FIRST SERVICE (KGS)				
AGE AT FIRST SERVICE (DAYS)				
HEAT DATES	DATE OF A.I/BULL SERVICE	PREG. DIAGNOSIS DATE	DATE TO DRY	DATE DUE TO CALF

COW ID #	

DATE OF SALE OR REMOVAL:	
REASON	
SALE WEIGHT	
SALE PRICE/LB.	
TOTAL VALUE	

AGE AT FIRST CALVING (DAYS)				
AGE AT CONCEPTION (DAYS)				
DATE CALVED	CALF SEX	CALF NO.	CALVING INTERNAL (DAYS)	NOTES

COW PRODUCTION HISTORY CARD

COW REGISTRATION #:			
DESCRIPTION (BREED/COLOR)			
COW'S SIRE		SIRE BREED:	
COW'S DAM		DAM BREED	
COW'S BIRTH DATE		WEANING WEIGHT	
PURCHASE DATE		PURCHASE PRICE	

BREEDING RECORD

WEIGHT AT FIRST SERVICE (KGS)			
AGE AT FIRST SERVICE (DAYS)			

HEAT DATES	DATE OF A.I/BULL SERVICE	PREG. DIAGNOSIS DATE	DATE TO DRY	DATE DUE TO CALF

COW ID #	

DATE OF SALE OR REMOVAL:	
REASON	
SALE WEIGHT	
SALE PRICE/LB.	
TOTAL VALUE	

AGE AT FIRST CALVING (DAYS)				
AGE AT CONCEPTION (DAYS)				
DATE CALVED	CALF SEX	CALF NO.	CALVING INTERNAL (DAYS)	NOTES

COW PRODUCTION HISTORY CARD

COW REGISTRATION #:			
DESCRIPTION (BREED/COLOR)			
COW'S SIRE		SIRE BREED:	
COW'S DAM		DAM BREED	
COW'S BIRTH DATE		WEANING WEIGHT	
PURCHASE DATE		PURCHASE PRICE	

BREEDING RECORD

WEIGHT AT FIRST SERVICE (KGS)			
AGE AT FIRST SERVICE (DAYS)			

HEAT DATES	DATE OF A.I/BULL SERVICE	PREG. DIAGNOSIS DATE	DATE TO DRY	DATE DUE TO CALF

COW ID #	

DATE OF SALE OR REMOVAL:	
REASON	
SALE WEIGHT	
SALE PRICE/LB.	
TOTAL VALUE	

AGE AT FIRST CALVING (DAYS)				
AGE AT CONCEPTION (DAYS)				
DATE CALVED	CALF SEX	CALF NO.	CALVING INTERNAL (DAYS)	NOTES

COW PRODUCTION HISTORY CARD

COW REGISTRATION #:			
DESCRIPTION (BREED/COLOR)			
COW'S SIRE		SIRE BREED:	
COW'S DAM		DAM BREED	
COW'S BIRTH DATE		WEANING WEIGHT	
PURCHASE DATE		PURCHASE PRICE	

BREEDING RECORD

WEIGHT AT FIRST SERVICE (KGS)				
AGE AT FIRST SERVICE (DAYS)				
HEAT DATES	DATE OF A.I/BULL SERVICE	PREG. DIAGNOSIS DATE	DATE TO DRY	DATE DUE TO CALF

COW ID #	

DATE OF SALE OR REMOVAL:	
REASON	
SALE WEIGHT	
SALE PRICE/LB.	
TOTAL VALUE	

AGE AT FIRST CALVING (DAYS)				
AGE AT CONCEPTION (DAYS)				
DATE CALVED	CALF SEX	CALF NO.	CALVING INTERNAL (DAYS)	NOTES

COW PRODUCTION HISTORY CARD

COW REGISTRATION #:			
DESCRIPTION (BREED/COLOR)			
COW'S SIRE		SIRE BREED:	
COW'S DAM		DAM BREED	
COW'S BIRTH DATE		WEANING WEIGHT	
PURCHASE DATE		PURCHASE PRICE	

BREEDING RECORD

WEIGHT AT FIRST SERVICE (KGS)				
AGE AT FIRST SERVICE (DAYS)				
HEAT DATES	DATE OF A.I/BULL SERVICE	PREG. DIAGNOSIS DATE	DATE TO DRY	DATE DUE TO CALF

COW ID

DATE OF SALE OR REMOVAL:	
REASON	
SALE WEIGHT	
SALE PRICE/LB.	
TOTAL VALUE	

AGE AT FIRST CALVING (DAYS)				
AGE AT CONCEPTION (DAYS)				
DATE CALVED	CALF SEX	CALF NO.	CALVING INTERNAL (DAYS)	NOTES

COW PRODUCTION HISTORY CARD

COW REGISTRATION #:	
DESCRIPTION (BREED/COLOR)	

COW'S SIRE		SIRE BREED:	
COW'S DAM		DAM BREED	
COW'S BIRTH DATE		WEANING WEIGHT	
PURCHASE DATE		PURCHASE PRICE	

BREEDING RECORD

WEIGHT AT FIRST SERVICE (KGS)			
AGE AT FIRST SERVICE (DAYS)			

HEAT DATES	DATE OF A.I/BULL SERVICE	PREG. DIAGNOSIS DATE	DATE TO DRY	DATE DUE TO CALF

COW ID

DATE OF SALE OR REMOVAL:	
REASON	
SALE WEIGHT	
SALE PRICE/LB.	
TOTAL VALUE	

AGE AT FIRST CALVING (DAYS)				
AGE AT CONCEPTION (DAYS)				
DATE CALVED	CALF SEX	CALF NO.	CALVING INTERNAL (DAYS)	NOTES

COW PRODUCTION HISTORY CARD

COW REGISTRATION #:			
DESCRIPTION (BREED/COLOR)			
COW'S SIRE		SIRE BREED:	
COW'S DAM		DAM BREED	
COW'S BIRTH DATE		WEANING WEIGHT	
PURCHASE DATE		PURCHASE PRICE	

BREEDING RECORD

WEIGHT AT FIRST SERVICE (KGS)				
AGE AT FIRST SERVICE (DAYS)				
HEAT DATES	DATE OF A.I/BULL SERVICE	PREG. DIAGNOSIS DATE	DATE TO DRY	DATE DUE TO CALF

COW ID #	

DATE OF SALE OR REMOVAL:	
REASON	
SALE WEIGHT	
SALE PRICE/LB.	
TOTAL VALUE	

AGE AT FIRST CALVING (DAYS)				
AGE AT CONCEPTION (DAYS)				
DATE CALVED	CALF SEX	CALF NO.	CALVING INTERNAL (DAYS)	NOTES

COW PRODUCTION HISTORY CARD

COW REGISTRATION #:			
DESCRIPTION (BREED/COLOR)			
COW'S SIRE		SIRE BREED:	
COW'S DAM		DAM BREED	
COW'S BIRTH DATE		WEANING WEIGHT	
PURCHASE DATE		PURCHASE PRICE	

BREEDING RECORD

WEIGHT AT FIRST SERVICE (KGS)			
AGE AT FIRST SERVICE (DAYS)			

HEAT DATES	DATE OF A.I/BULL SERVICE	PREG. DIAGNOSIS DATE	DATE TO DRY	DATE DUE TO CALF

COW ID #	

DATE OF SALE OR REMOVAL:	
REASON	
SALE WEIGHT	
SALE PRICE/LB.	
TOTAL VALUE	

AGE AT FIRST CALVING (DAYS)				
AGE AT CONCEPTION (DAYS)				
DATE CALVED	CALF SEX	CALF NO.	CALVING INTERNAL (DAYS)	NOTES

COW PRODUCTION HISTORY CARD

COW REGISTRATION #:	
DESCRIPTION (BREED/COLOR)	

COW'S SIRE		SIRE BREED:	
COW'S DAM		DAM BREED	
COW'S BIRTH DATE		WEANING WEIGHT	
PURCHASE DATE		PURCHASE PRICE	

BREEDING RECORD

WEIGHT AT FIRST SERVICE (KGS)			

AGE AT FIRST SERVICE (DAYS)			

HEAT DATES	DATE OF A.I/BULL SERVICE	PREG. DIAGNOSIS DATE	DATE TO DRY	DATE DUE TO CALF

COW ID #	

DATE OF SALE OR REMOVAL:	
REASON	
SALE WEIGHT	
SALE PRICE/LB.	
TOTAL VALUE	

AGE AT FIRST CALVING (DAYS)				
AGE AT CONCEPTION (DAYS)				
DATE CALVED	CALF SEX	CALF NO.	CALVING INTERNAL (DAYS)	NOTES

COW PRODUCTION HISTORY CARD

COW REGISTRATION #:			
DESCRIPTION (BREED/COLOR)			
COW'S SIRE		SIRE BREED:	
COW'S DAM		DAM BREED	
COW'S BIRTH DATE		WEANING WEIGHT	
PURCHASE DATE		PURCHASE PRICE	

BREEDING RECORD

WEIGHT AT FIRST SERVICE (KGS)			
AGE AT FIRST SERVICE (DAYS)			

HEAT DATES	DATE OF A.I/BULL SERVICE	PREG. DIAGNOSIS DATE	DATE TO DRY	DATE DUE TO CALF

COW ID

DATE OF SALE OR REMOVAL:	
REASON	
SALE WEIGHT	
SALE PRICE/LB.	
TOTAL VALUE	

AGE AT FIRST CALVING (DAYS)				
AGE AT CONCEPTION (DAYS)				
DATE CALVED	CALF SEX	CALF NO.	CALVING INTERNAL (DAYS)	NOTES

COW PRODUCTION HISTORY CARD

COW REGISTRATION #:			
DESCRIPTION (BREED/COLOR)			
COW'S SIRE		SIRE BREED:	
COW'S DAM		DAM BREED	
COW'S BIRTH DATE		WEANING WEIGHT	
PURCHASE DATE		PURCHASE PRICE	

BREEDING RECORD

WEIGHT AT FIRST SERVICE (KGS)			
AGE AT FIRST SERVICE (DAYS)			

HEAT DATES	DATE OF A.I/BULL SERVICE	PREG. DIAGNOSIS DATE	DATE TO DRY	DATE DUE TO CALF

COW ID #	

DATE OF SALE OR REMOVAL:	
REASON	
SALE WEIGHT	
SALE PRICE/L.B.	
TOTAL VALUE	

AGE AT FIRST CALVING (DAYS)				
AGE AT CONCEPTION (DAYS)				
DATE CALVED	CALF SEX	CALF NO.	CALVING INTERNAL (DAYS)	NOTES

COW PRODUCTION HISTORY CARD

COW REGISTRATION #:			
DESCRIPTION (BREED/COLOR)			
COW'S SIRE		SIRE BREED:	
COW'S DAM		DAM BREED	
COW'S BIRTH DATE		WEANING WEIGHT	
PURCHASE DATE		PURCHASE PRICE	

BREEDING RECORD

WEIGHT AT FIRST SERVICE (KGS)			
AGE AT FIRST SERVICE (DAYS)			
HEAT DATES	DATE OF A.I/BULL SERVICE	PREG. DIAGNOSIS DATE	DATE TO DRY	DATE DUE TO CALF

COW ID #	

DATE OF SALE OR REMOVAL:	
REASON	
SALE WEIGHT	
SALE PRICE/LB.	
TOTAL VALUE	

AGE AT FIRST CALVING (DAYS)				
AGE AT CONCEPTION (DAYS)				
DATE CALVED	CALF SEX	CALF NO.	CALVING INTERNAL (DAYS)	NOTES

COW PRODUCTION HISTORY CARD

COW REGISTRATION #:			
DESCRIPTION (BREED/COLOR)			
COW'S SIRE		SIRE BREED:	
COW'S DAM		DAM BREED	
COW'S BIRTH DATE		WEANING WEIGHT	
PURCHASE DATE		PURCHASE PRICE	

BREEDING RECORD

WEIGHT AT FIRST SERVICE (KGS)				
AGE AT FIRST SERVICE (DAYS)				
HEAT DATES	DATE OF A.I/BULL SERVICE	PREG. DIAGNOSIS DATE	DATE TO DRY	DATE DUE TO CALF

COW ID #	

DATE OF SALE OR REMOVAL:	
REASON	
SALE WEIGHT	
SALE PRICE/LB.	
TOTAL VALUE	

AGE AT FIRST CALVING (DAYS)				
AGE AT CONCEPTION (DAYS)				
DATE CALVED	CALF SEX	CALF NO.	CALVING INTERNAL (DAYS)	NOTES

COW PRODUCTION HISTORY CARD

COW REGISTRATION #:				
DESCRIPTION (BREED/COLOR)				
COW'S SIRE		SIRE BREED:		
COW'S DAM		DAM BREED		
COW'S BIRTH DATE		WEANING WEIGHT		
PURCHASE DATE		PURCHASE PRICE		

BREEDING RECORD

WEIGHT AT FIRST SERVICE (KGS)				
AGE AT FIRST SERVICE (DAYS)				
HEAT DATES	DATE OF A.I/BULL SERVICE	PREG. DIAGNOSIS DATE	DATE TO DRY	DATE DUE TO CALF

COW ID #	

DATE OF SALE OR REMOVAL:	
REASON	
SALE WEIGHT	
SALE PRICE/LB.	
TOTAL VALUE	

AGE AT FIRST CALVING (DAYS)				
AGE AT CONCEPTION (DAYS)				
DATE CALVED	CALF SEX	CALF NO.	CALVING INTERNAL (DAYS)	NOTES

COW PRODUCTION HISTORY CARD

COW REGISTRATION #:			
DESCRIPTION (BREED/COLOR)			
COW'S SIRE		SIRE BREED:	
COW'S DAM		DAM BREED	
COW'S BIRTH DATE		WEANING WEIGHT	
PURCHASE DATE		PURCHASE PRICE	

BREEDING RECORD

WEIGHT AT FIRST SERVICE (KGS)			
AGE AT FIRST SERVICE (DAYS)			

HEAT DATES	DATE OF A.I/BULL SERVICE	PREG. DIAGNOSIS DATE	DATE TO DRY	DATE DUE TO CALF

COW ID

DATE OF SALE OR REMOVAL:	
REASON	
SALE WEIGHT	
SALE PRICE/LB.	
TOTAL VALUE	

AGE AT FIRST CALVING (DAYS)				
AGE AT CONCEPTION (DAYS)				
DATE CALVED	CALF SEX	CALF NO.	CALVING INTERNAL (DAYS)	NOTES

COW PRODUCTION HISTORY CARD

COW REGISTRATION #:			
DESCRIPTION (BREED/COLOR)			
COW'S SIRE		SIRE BREED:	
COW'S DAM		DAM BREED	
COW'S BIRTH DATE		WEANING WEIGHT	
PURCHASE DATE		PURCHASE PRICE	

BREEDING RECORD

WEIGHT AT FIRST SERVICE (KGS)			
AGE AT FIRST SERVICE (DAYS)			

HEAT DATES	DATE OF A.I/BULL SERVICE	PREG. DIAGNOSIS DATE	DATE TO DRY	DATE DUE TO CALF

COW ID

DATE OF SALE OR REMOVAL:	
REASON	
SALE WEIGHT	
SALE PRICE/LB.	
TOTAL VALUE	

AGE AT FIRST CALVING (DAYS)				
AGE AT CONCEPTION (DAYS)				
DATE CALVED	CALF SEX	CALF NO.	CALVING INTERNAL (DAYS)	NOTES

COW PRODUCTION HISTORY CARD

COW REGISTRATION #:			
DESCRIPTION (BREED/COLOR)			
COW'S SIRE		SIRE BREED:	
COW'S DAM		DAM BREED	
COW'S BIRTH DATE		WEANING WEIGHT	
PURCHASE DATE		PURCHASE PRICE	

BREEDING RECORD

WEIGHT AT FIRST SERVICE (KGS)				
AGE AT FIRST SERVICE (DAYS)				
HEAT DATES	DATE OF A.I/BULL SERVICE	PREG. DIAGNOSIS DATE	DATE TO DRY	DATE DUE TO CALF

COW ID #	

DATE OF SALE OR REMOVAL:	
REASON	
SALE WEIGHT	
SALE PRICE/LB.	
TOTAL VALUE	

AGE AT FIRST CALVING (DAYS)				
AGE AT CONCEPTION (DAYS)				
DATE CALVED	CALF SEX	CALF NO.	CALVING INTERNAL (DAYS)	NOTES

COW PRODUCTION HISTORY CARD

COW REGISTRATION #:			
DESCRIPTION (BREED/COLOR)			
COW'S SIRE		SIRE BREED:	
COW'S DAM		DAM BREED	
COW'S BIRTH DATE		WEANING WEIGHT	
PURCHASE DATE		PURCHASE PRICE	

BREEDING RECORD

WEIGHT AT FIRST SERVICE (KGS)				
AGE AT FIRST SERVICE (DAYS)				
HEAT DATES	DATE OF A.I/BULL SERVICE	PREG. DIAGNOSIS DATE	DATE TO DRY	DATE DUE TO CALF

COW ID

DATE OF SALE OR REMOVAL:	
REASON	
SALE WEIGHT	
SALE PRICE/LB.	
TOTAL VALUE	

AGE AT FIRST CALVING (DAYS)				
AGE AT CONCEPTION (DAYS)				
DATE CALVED	CALF SEX	CALF NO.	CALVING INTERNAL (DAYS)	NOTES

COW PRODUCTION HISTORY CARD

COW REGISTRATION #:			
DESCRIPTION (BREED/COLOR)			
COW'S SIRE		SIRE BREED:	
COW'S DAM		DAM BREED	
COW'S BIRTH DATE		WEANING WEIGHT	
PURCHASE DATE		PURCHASE PRICE	

BREEDING RECORD

WEIGHT AT FIRST SERVICE (KGS)				
AGE AT FIRST SERVICE (DAYS)				
HEAT DATES	DATE OF A.I/BULL SERVICE	PREG. DIAGNOSIS DATE	DATE TO DRY	DATE DUE TO CALF

COW ID

DATE OF SALE OR REMOVAL:	
REASON	
SALE WEIGHT	
SALE PRICE/LB.	
TOTAL VALUE	

AGE AT FIRST CALVING (DAYS)				
AGE AT CONCEPTION (DAYS)				
DATE CALVED	CALF SEX	CALF NO.	CALVING INTERNAL (DAYS)	NOTES

COW PRODUCTION HISTORY CARD

COW REGISTRATION #:	
DESCRIPTION (BREED/COLOR)	

COW'S SIRE		SIRE BREED:	
COW'S DAM		DAM BREED	
COW'S BIRTH DATE		WEANING WEIGHT	
PURCHASE DATE		PURCHASE PRICE	

BREEDING RECORD

WEIGHT AT FIRST SERVICE (KGS)			
AGE AT FIRST SERVICE (DAYS)			

HEAT DATES	DATE OF A.I/BULL SERVICE	PREG. DIAGNOSIS DATE	DATE TO DRY	DATE DUE TO CALF

COW ID #	

DATE OF SALE OR REMOVAL:	
REASON	
SALE WEIGHT	
SALE PRICE/LB.	
TOTAL VALUE	

AGE AT FIRST CALVING (DAYS)				
AGE AT CONCEPTION (DAYS)				
DATE CALVED	CALF SEX	CALF NO.	CALVING INTERNAL (DAYS)	NOTES

COW PRODUCTION HISTORY CARD

COW REGISTRATION #:			
DESCRIPTION (BREED/COLOR)			
COW'S SIRE		SIRE BREED:	
COW'S DAM		DAM BREED	
COW'S BIRTH DATE		WEANING WEIGHT	
PURCHASE DATE		PURCHASE PRICE	

BREEDING RECORD

WEIGHT AT FIRST SERVICE (KGS)			
AGE AT FIRST SERVICE (DAYS)			

HEAT DATES	DATE OF A.I/BULL SERVICE	PREG. DIAGNOSIS DATE	DATE TO DRY	DATE DUE TO CALF

COW ID

DATE OF SALE OR REMOVAL:	
REASON	
SALE WEIGHT	
SALE PRICE/LB.	
TOTAL VALUE	

AGE AT FIRST CALVING (DAYS)				
AGE AT CONCEPTION (DAYS)				
DATE CALVED	CALF SEX	CALF NO.	CALVING INTERNAL (DAYS)	NOTES

COW PRODUCTION HISTORY CARD

COW REGISTRATION #:			
DESCRIPTION (BREED/COLOR)			
COW'S SIRE		SIRE BREED:	
COW'S DAM		DAM BREED	
COW'S BIRTH DATE		WEANING WEIGHT	
PURCHASE DATE		PURCHASE PRICE	

BREEDING RECORD

WEIGHT AT FIRST SERVICE (KGS)			
AGE AT FIRST SERVICE (DAYS)			

HEAT DATES	DATE OF A.I/BULL SERVICE	PREG. DIAGNOSIS DATE	DATE TO DRY	DATE DUE TO CALF

COW ID

DATE OF SALE OR REMOVAL:	
REASON	
SALE WEIGHT	
SALE PRICE/LB.	
TOTAL VALUE	

	AGE AT FIRST CALVING (DAYS)			
	AGE AT CONCEPTION (DAYS)			
DATE CALVED	CALF SEX	CALF NO.	CALVING INTERNAL (DAYS)	NOTES

MEDICAL LOG

DATE	COW ID	MEDICATION	DOSAGE	DIAGNOSIS	NOTES

DEWORMING & IMMUNIZATIONS

DATE	COW ID	TYPE	DOSAGE	DUE NEXT	NOTES

MEDICAL LOG

DATE	COW ID	MEDICATION	DOSAGE	DIAGNOSIS	NOTES

DEWORMING & IMMUNIZATIONS

DATE	COW ID	TYPE	DOSAGE	DUE NEXT	NOTES

MEDICAL LOG

DATE	COW ID	MEDICATION	DOSAGE	DIAGNOSIS	NOTES

DEWORMING & IMMUNIZATIONS

DATE	COW ID	TYPE	DOSAGE	DUE NEXT	NOTES

MEDICAL LOG

DATE	COW ID	MEDICATION	DOSAGE	DIAGNOSIS	NOTES

DEWORMING & IMMUNIZATIONS

DATE	COW ID	TYPE	DOSAGE	DUE NEXT	NOTES

MEDICAL LOG

DATE	COW ID	MEDICATION	DOSAGE	DIAGNOSIS	NOTES

DEWORMING & IMMUNIZATIONS

DATE	COW ID	TYPE	DOSAGE	DUE NEXT	NOTES

CATTLE SALES RECORD

DATE	CLASS OF CATTLE	ID NUMBER	WEIGHT	PRICE PER POUND	TOTAL PRICE
TOTAL SALES					$

EXPENSE RECORDS

| EXPENSE ITEM | | | | | | |
DATE	DESCRIPTION	FEED	SUPPLIES	MEDICAL	OTHER	TOTAL COST
TOTAL EXPENSES						$
NET PROFIT OR LOSS (TOTAL INCOME MINUS TOTAL EXPENSES)						$

CATTLE SALES RECORD

DATE	CLASS OF CATTLE	ID NUMBER	WEIGHT	PRICE PER POUND	TOTAL PRICE
TOTAL SALES					$

EXPENSE RECORDS

| EXPENSE ITEM | | | | | | |
DATE	DESCRIPTION	FEED	SUPPLIES	MEDICAL	OTHER	TOTAL COST
TOTAL EXPENSES						$
NET PROFIT OR LOSS (TOTAL INCOME MINUS TOTAL EXPENSES)						$

CATTLE SALES RECORD

DATE	CLASS OF CATTLE	ID NUMBER	WEIGHT	PRICE PER POUND	TOTAL PRICE
TOTAL SALES					$

EXPENSE RECORDS

EXPENSE ITEM						
DATE	DESCRIPTION	FEED	SUPPLIES	MEDICAL	OTHER	TOTAL COST
TOTAL EXPENSES						$
NET PROFIT OR LOSS (TOTAL INCOME MINUS TOTAL EXPENSES)						$

CATTLE SALES RECORD

DATE	CLASS OF CATTLE	ID NUMBER	WEIGHT	PRICE PER POUND	TOTAL PRICE
TOTAL SALES					$

EXPENSE RECORDS

EXPENSE ITEM						
DATE	DESCRIPTION	FEED	SUPPLIES	MEDICAL	OTHER	TOTAL COST
TOTAL EXPENSES						$
NET PROFIT OR LOSS (TOTAL INCOME MINUS TOTAL EXPENSES)						$

CATTLE SALES RECORD

DATE	CLASS OF CATTLE	ID NUMBER	WEIGHT	PRICE PER POUND	TOTAL PRICE
TOTAL SALES					$

EXPENSE RECORDS

EXPENSE ITEM						
DATE	DESCRIPTION	FEED	SUPPLIES	MEDICAL	OTHER	TOTAL COST
TOTAL EXPENSES						$
NET PROFIT OR LOSS (TOTAL INCOME MINUS TOTAL EXPENSES)						$

NOTES

NOTES

NOTES

NOTES

NOTES

NOTES

www.ingramcontent.com/pod-product-compliance
Lightning Source LLC
Chambersburg PA
CBHW051756200326
41597CB00025B/4570